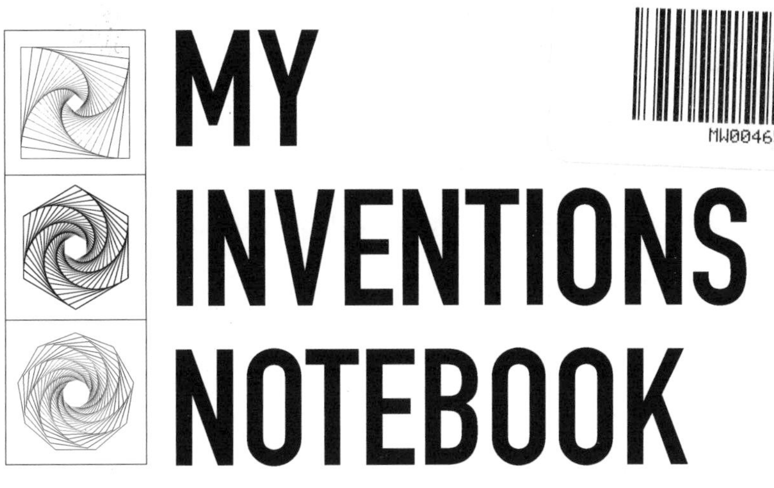

MY INVENTIONS NOTEBOOK

for the Aspiring Artist, Designer, Engineer, Maker, Creator, Inventor

This Book Belongs To:

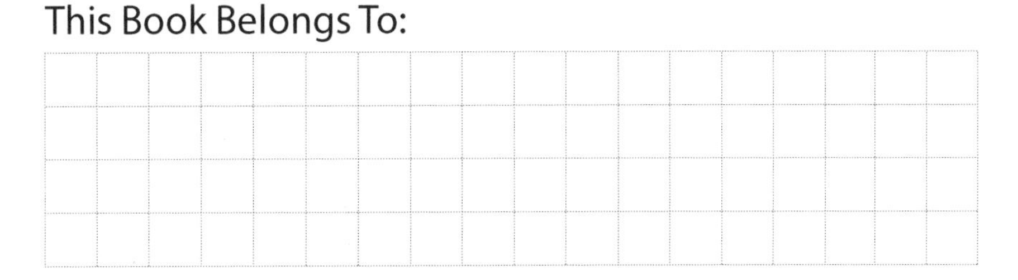

The Square/Isometric Combination Grid (Hybrid Grid) Design, Illustration, Layout and Text featured in this notebook is exclusive to the Createmplative series of utility journals for creatives.

ISBN-13: 978-1727348514
ISBN-10: 1727348516

The images referenced in the technical drawing explanations are from U.S. patent applications.
Source: United States Patent and Trademark Office, www.uspto.gov

Like an Inventor's Journal, this notebook is designed to help you keep track of the What, When, Where, Why and How of all your inventions. This is YOUR Notebook - it is all about documenting your creative path. Remember the #1 Rule of Creativity: There are No Rules - Mistakes are Impossible. **Just Start Creating!** If you do make a mistake – *that was a trick, there are no mistakes!* Don't erase it - embrace it. Sign and date it. Own it. Build upon everything you make. Continue to explore and develop your ideas as well as trying something completely new. **Keep Creating!** Your ideas and inventions can't change the world if you can't find them. Identify the awesomeness of each page by filling in the ...

TABLE OF CONTENTS

TABLE OF CONTENTS

TABLE OF CONTENTS

PAGE	TITLE/SUBJECT	CONTENT
67		
68		
69		
70		
71		
72		
73		
74		
75		
76		
77		
78		
79		
80		
81		
82		
83		
84		
85		
86		
87		
88		
89		
90		
91		
92		
93		
94		
95		
96		
97		
98		
99		
100		
101		
102		
103		

TITLE
1
SIGNATURE
DATE
WITNESS/REVIEWER
DATE

TITLE

SIGNATURE | DATE | WITNESS/REVIEWER | DATE

TITLE
SIGNATURE
DATE
WITNESS/REVIEWER
DATE

4
TITLE
SIGNATURE
DATE
WITNESS/REVIEWER
DATE
4

TITLE
5
SIGNATURE
DATE
WITNESS/REVIEWER
DATE
5

6
TITLE
SIGNATURE
DATE
WITNESS/REVIEWER
DATE
6

TITLE

SIGNATURE | DATE | WITNESS/REVIEWER | DATE

TITLE

SIGNATURE | DATE | WITNESS/REVIEWER | DATE

TITLE
SIGNATURE
DATE
WITNESS/REVIEWER
DATE
9

10
TITLE
SIGNATURE
DATE
WITNESS/REVIEWER
DATE
10

TITLE
11
SIGNATURE
DATE
WITNESS/REVIEWER
DATE
11

12
TITLE
SIGNATURE
DATE
WITNESS/REVIEWER
DATE
12

TITLE

SIGNATURE | DATE | WITNESS/REVIEWER | DATE

TITLE

SIGNATURE | DATE | WITNESS/REVIEWER | DATE

TITLE

SIGNATURE | DATE | WITNESS/REVIEWER | DATE

TITLE

SIGNATURE	DATE	WITNESS/REVIEWER	DATE

TITLE
SIGNATURE
DATE
WITNESS/REVIEWER
DATE
17

18
TITLE
SIGNATURE
DATE
WITNESS/REVIEWER
DATE
18

TITLE

SIGNATURE | DATE | WITNESS/REVIEWER | DATE

TITLE

SIGNATURE | DATE | WITNESS/REVIEWER | DATE

TITLE
21
SIGNATURE
DATE
WITNESS/REVIEWER
DATE
21

TITLE

SIGNATURE | DATE | WITNESS/REVIEWER | DATE

TITLE

SIGNATURE | DATE | WITNESS/REVIEWER | DATE

TITLE

SIGNATURE

DATE

WITNESS/REVIEWER

DATE

TITLE
25
SIGNATURE
DATE
WITNESS/REVIEWER
DATE
25

TITLE
SIGNATURE
DATE
WITNESS/REVIEWER
DATE
26

TITLE

SIGNATURE | DATE | WITNESS/REVIEWER | DATE

28
TITLE
SIGNATURE
DATE
WITNESS/REVIEWER
DATE
28

TITLE
SIGNATURE
DATE
WITNESS/REVIEWER
DATE
29

TITLE

SIGNATURE | DATE | WITNESS/REVIEWER | DATE

TITLE
31
SIGNATURE
DATE
WITNESS/REVIEWER
DATE
31

TITLE

SIGNATURE | DATE | WITNESS/REVIEWER | DATE

TITLE
SIGNATURE
DATE
WITNESS/REVIEWER
DATE
33

TITLE
SIGNATURE
DATE
WITNESS/REVIEWER
DATE

TITLE

SIGNATURE | DATE | WITNESS/REVIEWER | DATE

TITLE

SIGNATURE | DATE | WITNESS/REVIEWER | DATE

TITLE

SIGNATURE | DATE | WITNESS/REVIEWER | DATE

38
TITLE
SIGNATURE
DATE
WITNESS/REVIEWER
DATE
38

TITLE

SIGNATURE | DATE | WITNESS/REVIEWER | DATE

TITLE

SIGNATURE | DATE | WITNESS/REVIEWER | DATE

TITLE

SIGNATURE	DATE	WITNESS/REVIEWER	DATE

TITLE

SIGNATURE | DATE | WITNESS/REVIEWER | DATE

TITLE
43
SIGNATURE
DATE
WITNESS/REVIEWER
DATE
43

TITLE

SIGNATURE | DATE | WITNESS/REVIEWER | DATE

TITLE

SIGNATURE | DATE | WITNESS/REVIEWER | DATE

TITLE

SIGNATURE | DATE | WITNESS/REVIEWER | DATE

TITLE

SIGNATURE | DATE | WITNESS/REVIEWER | DATE

48
TITLE
48
SIGNATURE
DATE
WITNESS/REVIEWER
DATE

TITLE
49
SIGNATURE
DATE
WITNESS/REVIEWER
DATE
49

TITLE

SIGNATURE | DATE | WITNESS/REVIEWER | DATE

TITLE
51
SIGNATURE
DATE
WITNESS/REVIEWER
DATE
51

TITLE
SIGNATURE
DATE
WITNESS/REVIEWER
DATE
52

TITLE

SIGNATURE | DATE | WITNESS/REVIEWER | DATE

TITLE

SIGNATURE | DATE | WITNESS/REVIEWER | DATE

TITLE
SIGNATURE
DATE
WITNESS/REVIEWER
DATE
55

56
TITLE
SIGNATURE
DATE
WITNESS/REVIEWER
DATE
56

TITLE

SIGNATURE | DATE | WITNESS/REVIEWER | DATE

TITLE
SIGNATURE
DATE
WITNESS/REVIEWER
DATE
58

TITLE
SIGNATURE
DATE
WITNESS/REVIEWER
DATE
59

60
TITLE
SIGNATURE
DATE
WITNESS/REVIEWER
DATE
60

TITLE

SIGNATURE | DATE | WITNESS/REVIEWER | DATE

TITLE

SIGNATURE | DATE | WITNESS/REVIEWER | DATE

TITLE

SIGNATURE

DATE

WITNESS/REVIEWER

DATE

TITLE

SIGNATURE

DATE

WITNESS/REVIEWER

DATE

TITLE
SIGNATURE
DATE
WITNESS/REVIEWER
DATE
65

TITLE
SIGNATURE
DATE
WITNESS/REVIEWER
DATE
66

TITLE

SIGNATURE | DATE | WITNESS/REVIEWER | DATE

TITLE

SIGNATURE | DATE | WITNESS/REVIEWER | DATE

TITLE
69
SIGNATURE
DATE
WITNESS/REVIEWER
DATE
69

TITLE

SIGNATURE

DATE

WITNESS/REVIEWER

DATE

TITLE

SIGNATURE | DATE | WITNESS/REVIEWER | DATE

TITLE

SIGNATURE | DATE | WITNESS/REVIEWER | DATE

TITLE

SIGNATURE | DATE | WITNESS/REVIEWER | DATE

TITLE

SIGNATURE

DATE

WITNESS/REVIEWER

DATE

TITLE

SIGNATURE | DATE | WITNESS/REVIEWER | DATE

TITLE

SIGNATURE | DATE | WITNESS/REVIEWER | DATE

TITLE
SIGNATURE
DATE
WITNESS/REVIEWER
DATE
77

TITLE

SIGNATURE | DATE | WITNESS/REVIEWER | DATE

TITLE

SIGNATURE | DATE | WITNESS/REVIEWER | DATE

TITLE
SIGNATURE
DATE
WITNESS/REVIEWER
DATE
80

TITLE

SIGNATURE | DATE | WITNESS/REVIEWER | DATE

TITLE
SIGNATURE
DATE
WITNESS/REVIEWER
DATE
82

TITLE

SIGNATURE | DATE | WITNESS/REVIEWER | DATE

TITLE

SIGNATURE | DATE | WITNESS/REVIEWER | DATE

TITLE
85
SIGNATURE
DATE
WITNESS/REVIEWER
DATE
85

TITLE

SIGNATURE | DATE | WITNESS/REVIEWER | DATE

TITLE
SIGNATURE
DATE
WITNESS/REVIEWER
DATE
87

TITLE
SIGNATURE
DATE
WITNESS/REVIEWER
DATE

TITLE
SIGNATURE
DATE
WITNESS/REVIEWER
DATE
89

TITLE

SIGNATURE

DATE

WITNESS/REVIEWER

DATE

TITLE

SIGNATURE | DATE | WITNESS/REVIEWER | DATE

TITLE

SIGNATURE | DATE | WITNESS/REVIEWER | DATE

TITLE

SIGNATURE | DATE | WITNESS/REVIEWER | DATE

TITLE

SIGNATURE

DATE

WITNESS/REVIEWER

DATE

TITLE
95
SIGNATURE
DATE
WITNESS/REVIEWER
DATE
95

TITLE

SIGNATURE | DATE | WITNESS/REVIEWER | DATE

TITLE
SIGNATURE
DATE
WITNESS/REVIEWER
DATE

TITLE
SIGNATURE
DATE
WITNESS/REVIEWER
DATE

TITLE
99
SIGNATURE
DATE
WITNESS/REVIEWER
DATE
99

TITLE

SIGNATURE

DATE

WITNESS/REVIEWER

DATE

TITLE
101
SIGNATURE
DATE
WITNESS/REVIEWER
DATE
101

TITLE
SIGNATURE
DATE
WITNESS/REVIEWER
DATE
102

TITLE
SIGNATURE
DATE
WITNESS/REVIEWER
DATE
103

Made in the USA
Middletown, DE
15 March 2020